Made in the USA
Monee, IL
07 July 2026

56551540R00057